LUXURY SPACE MODEL GALLERY

奢华空间模型图库

建E网 编

辽宁科学技术出版社
·沈阳·

本书编委会

主编：陆 晏
编委：涂明成　王幸生　王銮甲　郭　凯　张华东
　　　杨　璨　孙　丽　陈繁力　何世洲　喻　明
　　　邹　刚　王廷和　李常春　刘湘龙　尹　亮
　　　梁建辉　李　健　常　宁　将贵宽　刘飞达
　　　孙传传　孙俭鑫　宋智红　张　伟　王洪波

图书在版编目（CIP）数据

奢华空间模型图库 / 建E网编. —沈阳：辽宁科学技术出版社，2013.5
 ISBN 978-7-5381-7971-2

Ⅰ.①奢… Ⅱ.①建… Ⅲ.①住宅－室内装修－建筑设计－图集 Ⅳ.① TU241-64

中国版本图书馆CIP数据核字（2013）第054762号

出版发行：辽宁科学技术出版社
　　　　　（地址：沈阳市和平区十一纬路29号　邮编：110003）
印　刷　者：恒美印务（广州）有限公司
经　销　者：各地新华书店
幅面尺寸：230mm×300mm
印　　张：32
插　　页：4
字　　数：300千字
印　　数：1~2000
出版时间：2013年5月第1版
印刷时间：2013年5月第1次印刷
责任编辑：郭媛媛
翻　　译：于　芳
封面设计：唐一文
版式设计：唐一文
责任校对：李　霞

书　　号：ISBN 978-7-5381-7971-2
定　　价：298.00元（随书附赠4张DVD光盘）

联系电话：024-23284356　13591655798
邮购热线：024-23284502
E-mail:purple6688@126.com
http://www.lnkj.com.cn

【序言】
PREFACE

经过近一年的酝酿，在众多读者与网友的期待下，我们与建E网合作推出了以"奢华"为主题的《奢华空间模型图库》。本书旨在献给室内设计师一个高端的模型图库，无论从设计风格，还是室内设施的布置，都散发着华美的气质、高端的品位。通过阅读本书，不仅能够获取更多的空间模型，还能感受到奢华空间带给人们的艺术魅力。

本书涉及项目319套，场景模型417个，效果图570张。主要包含了住宅、样板间、别墅豪宅等空间场景模型。模型3DMAX版本包括：269套2009版模型，11套2010版模型，7套2011版模型，26套2012版模型，6套9.0版本模型。所有模型经作者检查均含材质贴图、灯光参数等。本书内容丰富、实用性强，极具参考与收藏价值。

在此感谢建E网工作人员给予本书的大力支持。建E网自2008年正式上线运营以来，已经拥有注册设计师会员近100万名。收录高品质原创模型作品数万件，并保持每天更新。建E网是目前国内第一的原创室内设计模型发布与下载平台。网站的长期关注者和VIP会员多为国内一线高端设计公司与设计师。它创造性地整合了原创模型、设计店、设计资讯、设计行业招聘、软装材料等版块，成为国内室内设计最具综合实力的网站之一。建E网的资源已经为国内外广大设计师所熟悉和喜爱，大约500万名设计师正在使用建E网的资源。建E网致力于为建筑、室内、家具设计师提供高品质、高精度的3D模型，为中国室内设计师构建一个交流的平台。

Today "The Luxury Space Model Anthology" is published with full expectation from authors and netizens, which is a book focuses on luxury style. It has been edited by justeasy.cn net and our company together since one year ago. This book aims to help more interior designers to acquire more top-grade models. In this book, it emerges elegant and graceful taste of the design style and the home structure arranging. You will not only get many good space models but also experience the artistic charm of the luxury style.

This anthology consists of 319 projects, 417 scene models and 570 effect drawings. And mainly collects the space scene models on buildings, prototype rooms and villas. While the model 3DMAX editions contain 269 models made up of the 3DMAX 2009; 11 models made up of the 3DMAX 2010; 7 models made up of the 3DMAX 2011; 26 models made up of the 3DMAX 2012; and 6 models made up of the 3DMAX 9.0. All models in the book offer the material quality picture and the light data. All above advantages make this book has wide application and large collection value.

Here we thank for the support from the justeasy.cn net which has possessed one million members of registered designers during 5 years development. You can find countless latest and top models on this net. And the net supplies the models are original and high quality. As so far, the justeasy.cn net is the No.1 in the area of interior design model's publish and download in China. The fans and members of justeasy.cn net mostly are the domestic famous decorating design companies and designers. This net contributes to the creation of combination the original models, design shops, design information, professional recruitment and decorating materials. So it has become the most competitive design web in national-wide. Besides, the resources on the net have already been confirmed and rewarded by all the domestic and international designers. According to the statistics, there are more than 5 million designers using the resources on the justeasy.cn net. And the justeasy.cn net is devoting to offer the high quality and accuracy 3D models on building, architecture and furniture for the designers. The goal of the net is to shape a platform for communication of the domestic decorating designers.

[目录]

欧式风格 —————————————— 006

现代风格 —————————————— 060

中式风格 —————————————— 110

新古典风格 ————————————— 172

美式风格 —————————————— 212

地中海风格 ————————————— 242

CONTENTS

EUROPE STYLE —————————— 006

MODERN STYLE —————————— 060

CHINESE STYLE —————————— 110

NEW CLASSICAL STYLE ——————— 172

AMERICAN STYLE —————————— 212

MEDITERRANEAN STYLE ——————— 242

LUXURY SPACE MODEL GALLERY

欧式风格

EUROPE STYLE

LIFE IN CLASSICAL EUROPE
古典欧式的新生命力

MAKING EVERYTHING AT EASE
让一切回归安然

SEEKING AND SEARCHING
寻寻觅觅

TELLING AND LISTENING
娓娓道来，侧耳倾听

HISTORICAL ATMOSPHERE
复古格调

A030302A 盘

A030301A 盘

A140202A 盘

SILENCE LUXURY ELEMENTS
元素呈现低调奢华

ELEGANT AND DISTINGUISHED
高雅尊贵

CHARMING ORIGINATES FROM CULTURE
魅力来源于文化

A190902A

A192302A

A201501A

A201101A

GRACEFUL AND COMFORTABLE
优雅舒适

THE CLASSICAL ILLUSTRATING THE MODERN BEAUTY
糅古释今

LUXURY AND ELEGANCE
奢华高雅

A133101A

A100102A

A121503A 盘

A121506A 盘

A121501A 盘

A143001A

A160501A

A190801A

A171801A 盘

A191801A 盘

A191302A 盘

UNIQUE CHARACTERISTIC
独具特色

THE GENTLE FUN
柔和的趣味

LUXURY SPACE MODEL GALLERY

现代风格
MODERN STYLE

THE BEAUTY WITH CONCISE AND GRAND
简约大气之美

THE ELEGANT ACCOMPANY QUIETLY
淡雅安然相伴

CLEAN AND BRIGHT
干净明快

THE WONDER BEYOND WORDS EXPRESSING
美妙之处溢于言表

UNDISTURBED ENVIRONMENT,

静谧的环境，诗意生活中的都市恋曲

LOVE SONG IN PEOTRY LIFE

FASHIONAL LUXURY, SIMPLY CLASSIC

时尚的奢华，简约的古典

VOGUE AND FREEDOM
时尚且自由

B133301B 盘

B133401B 盘

PURE SPACE WITH BLACK, WHITE AND GRAY
黑、白、灰的纯净世界

THE PURITY OF ARTS
纯粹的艺术

NOBLENESS AND ELEGANCE
高贵优雅

BEAUTY OF MAGNIFICENCE
恢弘气度之美

WONDERFUL TECHNIQUES
完美的技术

APPRECIABLE ATMOSPHERE
赏析格调

THE UNIQUE AND SIMPLE DESIGN ELEMENT
独一无二的简约设计元素

THE TRADITIONAL AND BROAD STRUCTURE
中规中矩的宽阔格局

LUXURY IMAGE
奢华印象

B202003B 盘

B202002B 盘

B202001B 盘

B202802B

B202801B

B202803B

LUXURY SPACE MODEL GALLERY

中式风格
CHINESE STYLE

BRIGHT WINTERSWEET
傲艳的蜡梅

AUTHENTIC CHINESE STYLE
纯正的中国格调

CULTURE INHERITING THE CHARM
文化传承风韵

INTOXICATED IN THE HISTORY RIVER
沉醉于岁月的长河

MYSTERIOUS CHARM FROM THE CHINESE STYLE
中式味道的神秘意韵

GLAMOUR OF THE ORIENT
东方的韵味

TRANQUIL LIFE
禅意生活

AMAZING ETHICAL CHARM
最炫民族风

GLAMOUR WITH ANTIQUE FLAVOUR
古色古香的韵味

CAGING CHINESE STYLE
炫彩中国风

RETURN TO CHINESE CLASSICAL
回归中式古典

CHINESE ELEMENT
中国元素

C200101C 盘

C200601C 盘

C200401C 盘

C200501C 盘

LUXURY SPACE MODEL GALLERY

新古典风格

NEW CLASSICAL STYLE

CARVING THE NEOCLASSICAL ENGLAND STYLE
雕刻新古典英伦风

THE NEOCLASSICAL ATTRACT
新古典的诱惑

D121610B

D121609B

D132801B

D130801B 盘

D130901B 盘

MAKING ROMANTIC DREAM IN NEOCLASSICAL
在新古典中编织最浪漫的梦

D136701B 盘

D138001B 盘

D134401B 盘

D134402B 盘

NEOCLASSICAL BEAUTY SHAPED BY THE DETAILS
细节之处的新古典之美

THE MATING TRIP FROM HEART
心灵的相约之旅

THE BEAUTY WITH SPRING FLAVOR
春意盎然之美

D203201B

D201802B

D202701B

D221802B 盘

D141604B 盘

D135401B 盘

LUXURY SPACE MODEL GALLERY

美式风格

AMERICAN STYLE

INTOXICATED VISUAL BANQUET
令人陶醉的视觉盛宴

NEW VITALITY OF AMERICA STYLE
美式的新生命力

AMERICA STYLE HOUSE IN FAIRY KINGDOM
浸润在童话王国中的美式家

AMERICA CLASSIC WITH ROCK POWER
不可震撼的美式经典

NEW AMERICAN LIFE FULL OF FREEDOM
自由美式新生活

ELEGANCE COMBINED WITH DESIGN AND BEAUTY
设计与美的优雅结合

E121206D 盘

E131001D 盘

ENJOY THE TASTING STYLE
品位格调的赏析

PERFECT EMERGING
完美呈现

THE SENSE OF ALIENATION MADE BY THE CLASSICAL LUXURY
复古奢华的空间造成的疏离感

GRACEFUL LUXURY
奢华雅致

MAKING THE VOGUE TO BE BRIGHT AND COLORFUL

晕染了时尚界的明亮艳丽

E140804D 盘

E140805D 盘

E120608D 盘

FUSION OF CLASSICAL ELEMENTS
经典元素的融合

E171201D 盘

E172401D 盘

NPOURING THE HISTORY AND POWER OF TIME
注入了时间的沧桑和力度

SHAPING THE INTELLIGENT OF THE ROOM
衬托空间的灵动性

LUXURY SPACE MODEL GALLERY

地中海风格

MEDITERRANEAN STYLE

THE BLUE FLYING LIKE THE SEA
犹如海的蓝在飞扬

SWAYING WITH GRACEFUL WIND
清雅飘逸的风吹拂

【鸣谢】
THANKS

80设计涂明成

现任职务：80设计表现组长
工作经历：2007—2008年湖北美术学院3DMAX培训班指导老师！为即将毕业的学员进行一定的基础软件培训！2008—2009年蒙娜丽莎装饰有限公司设计师兼效果图表现人员！在做设计师的同时为公司的项目进行效果图表现，尽可能地表达出自己和客户所追求的氛围和效果！2009—2012年80设计工作室表现组长！至此已专注效果图研究5年，为了更好的效果、更高的层次，我会继续努力！我们的宗旨是"全心全意为您分忧，服务是本质，沟通是方式，务实是作风"。
座 右 铭：每一个成功者都在找原因，每一个失败者都在找借口！
电   话：15994287737 QQ：841241450（图为130101-138101）

深圳南辕图像王幸生

南辕图像于2010年成立于深圳．团队人员有着来自中国香港、深圳等多家设计公司工作经验。专致于酒店、酒吧、KTV、售楼处、样板房等视觉体现，本着精于品、锐于质的宗旨而不懈努力！
地   址：深圳市南山区沙河西路4811号深港花卉中心N-05
作品链接：http://spring2820.shijuew.com
电   话：13751076940 QQ：283273651（图为140101-143901）

王銮甲

电   话：13269262022 QQ：86945486（图为200101-203402）

武汉吉力室内设计

工作室成立于2009年，服务于地产商，设计公司，高端设计工作室。一贯专注于高端效果图表现，为客户提供高效优质的服务，完成了大量高品质作品，得到客户广泛认可！我们不断学习钻研国内外最新表现技术，力求达到：
　　模型精准，色彩和谐！
　　光影丰富，构图合理！
　　同样空间，别样效果！
　　极速出图，优质服务！
在表现领域不断拓展市场空间，与众多公司建立了良好的合作关系，赢得了客户的广泛好评！
郭凯 QQ：591562909
电   话：13554380007（图为160101-164603）

合肥一零二一装饰工程有限公司

张华东：毕业于阜阳师范学院美术系，专修室内设计专业。2006—2013年1月就职于上海统帅建筑装潢有限公司。2013年2月加入合肥一零二一装饰工程有限公司。本着"发现需求，满足需求"的设计理念，凭借一贯诚信和务实的工作作风，以专业设计、规范施工和用心服务，受到了客户的一致好评！代表作品楼盘：上海绿地威廉、长泰西郊别墅、东郊华庭别墅、万科白马花园、陆家嘴公寓等。

杨　璨：毕业于阜阳师范学院美术系，专修室内设计专业。2006年至今就职于合肥一零二一装饰工程有限公司，任运营总监兼首席设计师。代表作品楼盘：梦圆小区、海顿公馆、维多利亚别墅、绿地内森庄园别墅等。

张华东 QQ：214848692　电话：18226083344（图为190401-192501）

杨　璨 QQ：46493247　电话：15900649646

孙　丽 QQ：81101673　电话：13500776310（图为170101-172401）

陈繁力 QQ：157693859　电话：13732056080（图为220101-222001）

何世洲 QQ：451889830　电话：15055789058（图为120101-121506）

喻　明 QQ：229676743　电话：18627577317（图为180101-181201）

重庆蓝钻图文设计有限公司

邹　刚 QQ：309920732　电话：13996325089（图为070101-070701）

王廷和 QQ：361851341　电话：13613607744（图为050101-050701）

李常春 QQ：527269476　电话：15170883810（图为010102-010605）

刘湘龙 QQ：373600530　电话：15900663990（图为030101-030602）

尹　亮 QQ：14022178　电话：15982118212（图为080101-080603）

梁建辉 QQ：461532398　电话：15867127729（图为090101-090801）

李　健 QQ：350819249　电话：15996582793（图为210101-210603）

常　宁 QQ：897927926　电话：18809865502（图为100101-100601）

将贵宽 QQ：630176258　电话：13916875618（图为020101-020601）

刘飞达 QQ：381369655　电话：13560999901（图为110101-110403）

孙传传 QQ：252623503　电话：18210537768（图为150101-150701）

孙俭鑫 QQ：541300583　电话：13917320096（图为040101-040302）

宋智红 QQ：594383008　电话：13466362522（图为230101-230301）

张　伟 QQ：653408657　电话：15890720302（图为240101-240203）

王洪波 QQ：363789935　电话：15828232223（图为060203-060206）